The colorful stripes of the Grand Canyon show the many layers of rock.

12 ICONIC AMERICAN NATURAL WONDERS

BLACK RABBIT BOOKS

RACHEL GRACK

Table of Contents

Deep Colors
at the
Grand Canyon

The Grand Canyon is one of the Seven Wonders of the World. It stretches 277 miles (446 kilometers) across northern Arizona. At its widest, it spans 18 miles (29 km) across. It has steep sides. In some areas, the canyon is 1 mile (1.6 km) deep.

The Grand Canyon was formed by **erosion**. The canyon walls are made up of layers of lava and rock. Scientists believe the Colorado River began cutting through the layers of rock millions of years ago. Today, the river twists through the bottom of the canyon.

Native peoples of the Southwest have deep ties to the Grand Canyon. The land has been part of their culture for thousands of years. Today, the park shares boundaries with 11 native tribes. The canyon is also home to many plants and wildlife.

About 5 million people visit Grand Canyon National Park each year. They explore the North and South Rims. Many enjoy hiking along the South Rim. This is a popular **tourist** area. The North Rim is rougher and more secluded. It is also higher than the South side. People stand at the rim and look out across the canyon. They are amazed by the beautiful rocks.

Hiking in the Grand Canyon is hard work but the incredible view is worth it.

The canyon's rock seems to glow during Arizona sunsets.

Life Thrives in *Death Valley*

Death Valley is a desert in southeastern California. It is the hottest place in North America. Summer temperatures often reach 120 degrees Fahrenheit (49 degrees Celsius). The desert is very dry too. Most rain is blocked by the mountains in the west. Average rainfall is less than 2 inches (5.1 centimeters) a year. Some years, there is no rain at all.

Not all of Death Valley is desert, though. It has natural springs, wetlands, and salt lakes. There is wildlife too. Animals depend on these water sources. Coyotes, foxes, and bobcats hunt rabbits and rodents. Herds of bighorn sheep come down from the mountains. They graze on salt grass and desert brush. Lizards, snakes, and scorpions thrive in the dry landscape.

Coyotes are one of the most common animals found in Death Valley.

134.1 Hottest temperature in degrees Fahrenheit (56.7 °C) recorded at Death Valley, in 1913. That is a world record!

Death Valley National Park covers 3.4 million acres (1.4 million hectares). • It is the largest national park in the continental US. • More than 1 million people visit the park each year.

°F °C
120
100
80
60
40
50
40
30
20
10
0
10
20
30

Death Valley National Park is a popular tourist site. There are many landforms to see. The desert is dotted with volcanic craters. **Salt flats** spread across Badwater Basin. Other areas feature rippling sand dunes and **badlands**. People marvel at the "sailing stones." These shifting stones make paths in the sand. However, no one has seen them move. Death Valley offers incredible views of nature's handiwork.

A SPOOKY NAME

Death Valley holds plenty of beauty and wildlife. So, how did it get such a spooky name? In 1849, a group of settlers were heading to California. They crossed the valley. They ended up stranded there over the winter. They ran out of food and many of them died. One survivor gave the valley its grim name.

10

Everglades Overflow *with Wildlife*

3

The Everglades is a subtropical wetland. It is in southern Florida. It was formed by a shallow river. The water slowly flows from the Kissimmee River to Lake Okeechobee to Florida Bay. The waters pass through thick sawgrass and **mangrove** forests. The Everglades cover more than 1.5 million acres (607,028 ha). Changes in water depth and saltiness create different habitats. The Everglades contains nine **ecosystems**. It's the only place where both alligators and crocodiles live together.

Many types of plants and animals live throughout the Everglades. It is home to more than 800 **species** of wildlife. More than 360 kinds of birds gather in its marshes. Dozens of types of orchids grow there.

There are about 200,000 alligators in the Everglades.

Tourists can take kayaks through the mangrove swamps in the Everglades.

The mangroves are important habitats for manatees and wading birds. Many threatened and endangered animals depend on the Everglades. These include Florida panthers, American crocodiles, sea turtles, and manatees.

The Everglades is like no other place in the world. Its vast wilderness offers unique views of nature. People from around the world visit. They take guided boat tours. Some go canoeing. There are areas for hiking and biking too.

Florida panther

13

Mammoth Caves Lie Under *Kentucky*

4 Mammoth Cave is the longest cave system in the world. It lies under the hills of Kentucky. The national park covers about 80 square miles (207 square km) above ground. But no one knows how many miles (km) of caves are underground. There are five levels of passages. Over 400 miles (644 km) of caves have been explored so far. Many more wait to be discovered.

The cave system was formed by two layers of stone. The top layer has sandstone and shale. It acts like an umbrella over a lower layer of limestone. The top layer also has leaks that create sinkholes. Water drains from the surface in these places. The water erodes the limestone below, creating caves and caverns. The dripping water carries minerals that create rock formations. Groundwater also dissolves the limestone. This forms underground streams and rivers.

1790s

When Mammoth Cave was found by American settlers.

The caves are 54 °F (12 °C) year-round. • The deepest caves are 379 feet (116 m) underground. • There are 10 miles (16 km) of caves open to tourists.

Think About It

Caves can be dark, wet, and rocky. What types of gear would you need to explore a cave?

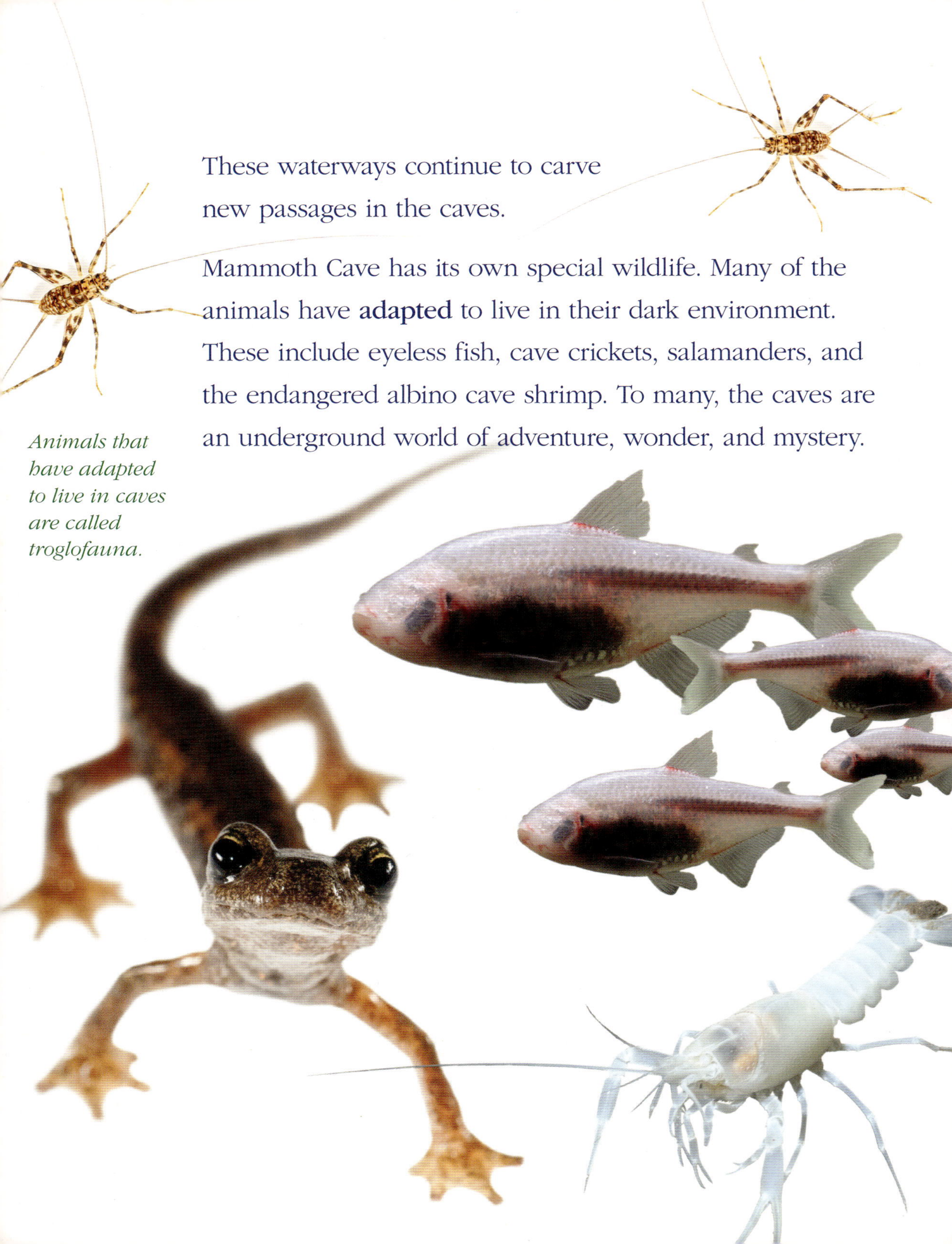

These waterways continue to carve new passages in the caves.

Mammoth Cave has its own special wildlife. Many of the animals have **adapted** to live in their dark environment. These include eyeless fish, cave crickets, salamanders, and the endangered albino cave shrimp. To many, the caves are an underground world of adventure, wonder, and mystery.

Animals that have adapted to live in caves are called troglofauna.

Big Trees Tower in *Giant Forest*

The Giant Forest is a **grove** of the largest trees in the world. Over 8,000 giant sequoias grow there. They stand between 250 and 300 feet (76–91 m) tall. That's almost as tall as the Statue of Liberty. Some trunks are over 30 feet (9.1 m) wide.

Sequoias only grow in certain parts of the Sierra Nevada Mountains in California. There are about 75 groves of them. The Giant Forest is one of the largest. It lies in the heart of Sequoia National Park. Many of the sequoias are over 2,000 years old.

The biggest tree in the Giant Forest is the General Sherman Tree. It stands 275 feet (84 m) tall and is 100 feet (31 m) around. It is around 2,500 years old. Other trees grow in the park too. There are redwoods, pines, fir, and cypress.

The Giant Forest is a popular
place to visit. People hike
between the towering trees.
Some trees have tunnels cut
through them. Trails and
roads pass right through
the trunks. People look tiny
next to the enormous trees.
A walk through the Giant
Forest is like a trip into
another world.

Some sequoias grow almost as tall as the Statue of Liberty.

3,400 Number of years a giant sequoia can live.
The bark on a giant sequoia is 24 inches (61 cm) thick. • General Sherman Tree's trunk weighs 626 tons (568 metric tons). • The Giant Forest is 1,800 acres (728 ha).
Sequoias have red-orange bark that stands out against the brown and green colors of other trees.

Think
About
It
The trees
in the
Giant Forest
can live
thousands
of years.
Why is it
important
to preserve
these
big trees?

Mississippi River Carves a Path *through America*

The Mississippi River is one of the most important waterways in the world. It travels 2,340 miles (3,766 km) across the United States. It flows from northern Minnesota to the Gulf of Mexico. With its **tributaries**, the Mississippi drains waterways in 31 states in the United States and 2 provinces in Canada. It is a key river for moving goods throughout the country. The river also serves as a **migration** path for many fish and birds.

The Upper Mississippi starts in Lake Itasca in Minnesota. It winds its way to the Ohio River in Illinois. Channels, locks, and dams help larger boats move freight. They carry goods from the Midwest to other parts of the country.

The Lower Mississippi starts at the mouth of the Ohio River. It flows freely without locks or dams. Its branching waterways form the largest **floodplain** in the United States.

The paddlefish is one of many species of fish found in the Mississippi River.

The floodplain covers more than 24 million acres (9.7 million ha). It spreads across parts of seven states, from Illinois to Louisiana. The floodwaters create rich farmland. Its wetlands are homes for migrating birds and waterfowl. And it is a major shipping system for America.

250 million

Number of tributaries and branches of the Mississippi River.

Ten states border the Mississippi River. • The Upper Mississippi has 29 locks and dams. • Over a year, 460 million tons (417 million mt) of freight are moved along the river.

The Mississippi River flows through downtown Memphis, Tennessee.

White pelicans live
along the upper
Mississippi River and
travel to the Gulf of
Mexico in the winter.

Denali Reaches *Great Heights*

7 Denali is the highest mountain in North America. Its peak rises to 20,310 feet (6,191 m). Denali is part of the Alaska Range of mountains. Native Alaskan people gave the mountain its name. It means "the high one."

Denali was formed by shifting plates in the Earth's crust. It is a giant wedge of granite. Its upper half is covered with snow and glaciers. The longest glacier is 44 miles (71 km) long. The plates under Denali are still moving. The peak grows by 0.04 inches (1 millimeter) every year.

Several hundred people try climbing the mountain each year. Its steep sides make it a difficult climb. The weather can get very cold too. Temperatures can drop to -75 °F (-59 °C) at night. Only about half of the climbers make it to the top.

Over 1,000 people climb
Denali every year.

11 Age of the youngest climber to reach
the top of Denali.

The first climbers reached the peak in 1913.
• Denali's peak is always covered by snow. • The
mountain was once called Mount McKinley. The
name was changed in 2015.

The mountain is part of Denali National Park. The park covers 6 million acres (2.4 million ha) of Alaskan wilderness. Only one road runs through the park. Most of the park is traveled only by bus. Wildlife includes bears, mountain goats, wolves, otters, and caribou.

Mountain goats have adapted to live in the cold, rocky terrain and high elevation.

Old Faithful Erupts *Like Clockwork*

Old Faithful is the most famous **geyser** in the world. It erupts every 60 to 90 minutes. Not all geysers are so regular. But that's how Old Faithful got its name. Old Faithful is in Yellowstone National Park in Wyoming. Yellowstone is one of the most active geyser areas in the world. It has more than 500 active geysers. They formed about 15,000 years ago.

8

Old Faithful is a cone geyser. These types of geysers are somewhat uncommon. They only form under special conditions. The ground contains tunnels lined with volcanic rock. These pipe-like holes can handle incredible heat and pressure Old Faithful shoots up to 8,400 gallons (31,798 liters) of sizzling hot water for two to five minutes.

Yellowstone is as famous for its wildlife, such as bison, as its geysers.

WHY GEYSERS SPOUT

Geysers spout when groundwater is heated by magma. They occur in places where volcanoes once existed. The water gets hotter and hotter. This causes steam to form in underground tunnels. Pressure builds under the geyser. Eventually, there is too much steam to hold. It releases in a sudden burst of steam and water.

Millions of people visit Old Faithful every year. They gather around the geyser and anxiously wait. Suddenly, a column of water bursts out of the ground. It shoots up to 140 feet (43 meters) into the sky. Steam billows out as it explodes. It is a stunning sight. People travel from around the world to get a glimpse of Old Faithful.

204 Temperature in degrees Fahrenheit (96 °C) of Old Faithful.

Old Faithful was discovered by American explorers in 1870. • There are more than 300 geysers in Yellowstone. • The other type of geyser is a fountain geyser.

A herd of bison grazes next to Old Faithful as it is about to erupt.

Niagara Falls Flows at *Incredible Rates*

9 Rushing water crashes down Niagara Falls. It is one of the most famous tourist sites in North America. Niagara Falls is created by the Niagara River. This river runs between the United States and Canada. It flows into a **gorge** between New York and Canada. There, the river splits into three waterfalls. Horseshoe Falls in Ontario, Canada, is the biggest. It is 2,200 feet (671 m) wide. It drops 188 feet (57 m) into the Lower Niagara River. American Falls and Bridal Veil Falls are in the United States. American Falls is 1,060 feet (323 m) across.

Niagara Falls is not the tallest waterfall in the world. But it has the highest flow rate. Over 750,000 gallons (2.8 million liters) of water spill over the falls every second. The amount of water flowing over is what makes it so spectacular. The waterfalls are used to produce electric power for both the United States and Canada.

Horseshoe Falls is more popular with tourists due to its impressive size.

Tourists can see both
falls at Prospect Point.

Niagara Falls was once a popular destination for newlyweds. Today, people from around the world visit its impressive beauty. The tumbling water causes clouds of mist. Boats also carry visitors through the churning waters below. People on board get soaked by the mist and hear the great roar of the water. It is a stunning sight up close.

1885 Year Niagara State Park was established.

It was the first US state park. • Four of the Great Lakes drain into the Niagara River. • Power plants by the falls use the rushing water to generate electricity.

A ferry boat brings visitors up close up to Horseshoe Falls.

Lava Flows Constantly at *Kilauea Volcano*

10

Kilauea is the most active volcano in the world. It lies on the Big Island of Hawaii in the Pacific Ocean. It is part of Hawaii Volcanoes National Park. Many people think of volcanoes as having steep sides. Kilauea looks different. It is a shield volcano. These have low slopes and dozens of craters.

Kilauea has been constantly erupting since 1983. Most eruptions are gentle lava flows. Each eruption, the lava flowed out and hardened. This made the volcano grow taller. Its **summit** once reached 4,090 feet (1,247 m). But in 2018, it collapsed. Huge clouds of ash darkened the sky. Bursts of lava shot 330 feet (101 m) into the air. A large lava flow spread over nearby neighborhoods. It destroyed hundreds of homes. A series of earthquakes followed.

1,900 Height in feet (580 m) of lava that shot out of Kilauea in 1959.

Kilauea has covered about 500 square miles (1,295 sq km) in lava since 1983. • There are viewing spots where visitors can see the active crater. • There are 6 active volcanoes in Hawaii.

Think About It Kilauea means "much spreading" in the Hawaiian language. How is this a fitting name for this volcano?

The volcano remained quiet until 2020. Then, small eruptions began again. They continue today. The flowing lava will slowly build up until the next big eruption. Kilauea is always changing and full of surprises. This adds to its wonder.

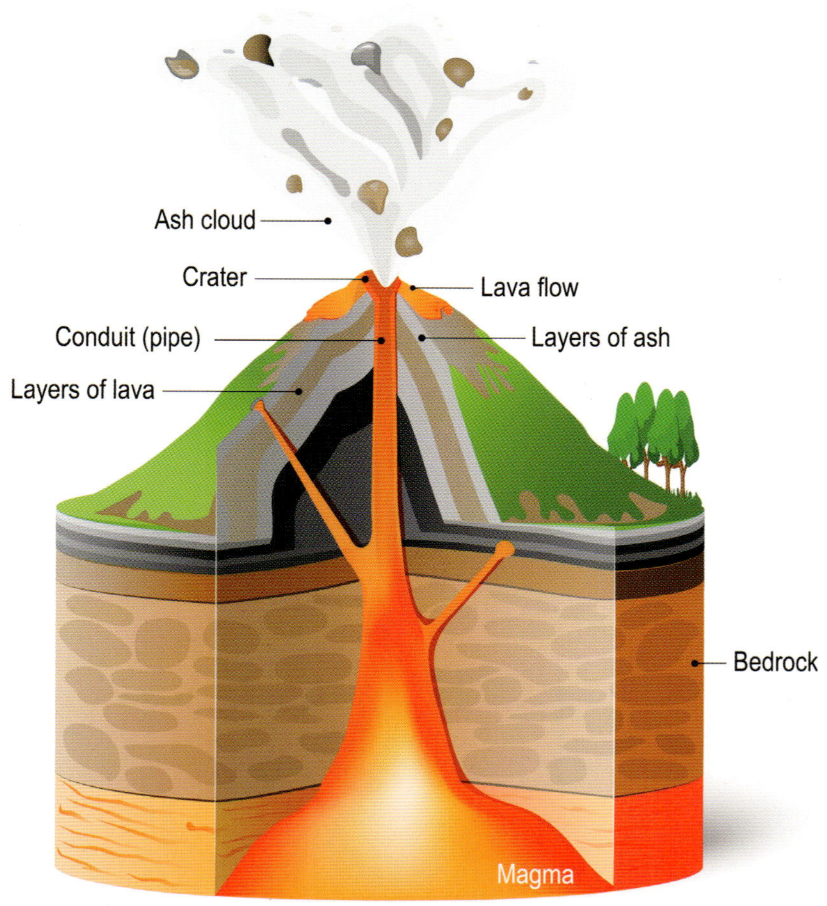

Crystal Clear Water at *Crater Lake*

Crater Lake in Oregon is the deepest US lake. It has an average depth of 1,943 feet (592 m). The lake holds 4.9 trillion gallons (18.6 trillion liters) of water.

Crater Lake began as Mount Mazama. It was a volcano in the Cascade Range. It stood about 12,000 feet (3,658 m) tall. About 7,700 years ago, it erupted. The center of the mountain collapsed. It left a deep caldera. Other eruptions followed. These formed Wizard Island in the middle of the crater. The volcano is no longer active.

Evaporation keeps Crater Lake at a certain level. No rivers feed the lake. It gets refilled by snow and rain in the spring. Crater Lake gets about 43 feet (13 m) of snow every year. It is one of snowiest places in the country. Due to its size, the lake rarely freezes over in the winter.

Crater Lake is known for its brilliant blue color. The water is very clear. Rainbow trout and salmon live in the lake. Rough-skinned salamanders called Mazama Newts also live there. People visit the lake throughout the year. It holds great beauty in both winter and summer.

THE OLD MAN

The Old Man is a 450-year-old log in Crater Lake. It was first spotted in 1896. The log floats straight up and down. It is 30 feet (9.1 m) long, but only 3 feet (0.9 m) stick out of the water. The Old Man bobs and drifts around the lake. People wonder how it got there and how long it will stay afloat.

286

Size of Crater Lake in square miles (741 square kilometers).

The highest point along the rim is 8,151 feet (2,484 m). • It has been 4,800 years since Mount Mazama's last eruption. • Thick moss covers the lake below the surface.

Rocky Mountains
Rise across
the West

12

The Rocky Mountains form the longest mountain range in North America. They stretch 3,000 miles (4,828 km). The mountains start in the center of New Mexico. They run northwest through Colorado, Utah, Wyoming, Montana, and Washington. The jagged peaks continue into Canada.

The Rocky Mountains include 100 mountain ranges. The US ranges are separated into four groups. The Northern Rockies are in Montana and northwest Idaho. The Middle Rockies span Wyoming, Utah, and southeast Idaho. The Southern Rockies lie mostly in Colorado and New Mexico. The Colorado Plateau covers the Four Corners of Utah, Colorado, New Mexico, and Arizona. The part of the mountains that spread into Canada are called the Canadian Rockies.

The Four Corners is marked with a plaque on the ground.

Moraine Lake in
Alberta, Canada, is
surrounded by the
Canadian Rockies.

The Rockies have high **elevations**. Many peaks reach over 13,000 feet (3,962 m). They also include valleys, rivers, and glaciers. The types of plants and wildlife differ throughout the range. Bears, mountain lions, and wolves live in pine, fir, and aspen forests. Bighorn sheep and deer graze in meadows and grasslands. There are snowcapped peaks, towering trees, and colorful wildflowers. These all make the Rockies one of the most majestic mountain ranges in the world.

Moraine Lake's
crystal clear
waters are famous
worldwide.

Where *in America?*

1. **Crater Lake**
Oregon

2. **Death Valley**
California

3. **Denali**
Alaska

4. **Everglades**
Florida

5. **Giant Forest**
California

6. **Grand Canyon**
Arizona

7. **Kilauea Volcano**
Hawaii

8. **Mammoth Cave**
Kentucky

9. **Mississippi River**
Central US

10. **Niagara Falls**
New York and Ontario

11. **Old Faithful**
Wyoming

12. **Rocky Mountains**
Western US and Canada

44

CANADA
Ontario
New York
Minnesota
Iowa
Missouri
Kentucky
Tennessee
Arkansas
Louisiana
Florida
9
9
10
9
9
8
9
9
9
4

Glossary

badlands
A dry area with few plants where weather has worn away rocks into strange shapes.

caldera
A large crater caused by a volcanic eruption.

ecosystem
A community of living things in one place.

elevation
The height of a place.

erosion
The slow wearing away of something by water or wind.

evaporation
The process of water turning from liquid into gas or vapor.

floodplain
An area of low, flat land along a stream or river that may flood.

geyser
A hole in the ground that shoots out hot water and steam.

gorge
A deep, narrow area between hills or mountains.

mangrove
A tropical tree that grows in swamps or shallow salt water.

migration
When animals move from one area to another at different times of the year.

salt flat
An area of flat land covered with a layer of salt.

species
A kind of animal or plant.

summit
The highest point of a mountain.

tourist
A person who travels for fun.

For More Information

Books

Cooke, Tim. *Giant's Causeway and Other Incredible Natural Wonders.* Minneapolis: Lerner Publications, 2024.

Frisch, Nate. *Rocky Mountain National Park.* Mankato, MN: Creative Education & Creative Paperbacks, 2025.

Gleason, Carrie. *You Are Here, United States.* New York: Crabtree Publishing, 2024.

Websites

Everglades Mountains and Valleys (videos)
home.nps.gov/ever/learn/photosmultimedia/
mountainsandvalleys.htm

Grand Canyon
kids.britannica.com/kids/article/Grand-Canyon/346126

Old Faithful Virtual Visitor Center
www.nps.gov/features/yell/ofvec/exhibits/index.htm

About the Author

Rachel Grack has been editing and writing children's books since 1999. She lives on a small ranch in the Arizona desert with her horse, Lady, and a bouncy goat named Pogo. Arizona offers countless natural features and historic sites. Rachel enjoys day trips exploring the beautiful landscape and learning more about Arizona's rich history.

Index

Copyright © 2025 Black Rabbit Books. All rights reserved. No part of this book may be reproduced in any form without written permission from the publisher. • Top Rank is an imprint of Black Rabbit Books. • Edited by Alissa Thielges | Designed by Danny Nanos • Photographs © Alamy Stock Photo/B. Mete Uz, 16–17, Darwin Wiggett, 32, Michael Weber, 24; Dreamstime/Bryan Busovicki, cover, 1, Designua, 36, Freerlaw, 11, Karen Foley, 29, Pavel Shlykov, 2, Rolffimages, 6, Svetlana Foote, 13; Shutterstock/Alexey Suloev, 4, Anthony Ricci, 2–3, 11, aslysun, 6, Belinda Pretorius, 48, BEST-BACKGROUNDS, 21, BestStockFoto, 14, Bill Pena, 5, Dan Olsen, 16, Dennis MacDonald, 33, Eric Isselee, 16, 27, 42, FedBul, 37–38, Federico.Crovetto, 16, Fernando Tatay, 20, FloridaStock, 41, Galyna Andrushko, 24–25, Gary Saxe, 19, Ghost Bear, 43, Happy Stock Photo, 33, Henryk Sadura, 22, Islamic Footage, 35, Jnjphotos, 27, 28, Jolygon, 46–47, Joy Sagar, 11, kavram, 8–9, KensCanning, 8, Ko Zatu, 15, Kriso, 15, Kurit afshen, 10, Legacy Images, 40, mariakray, 12, Matthew Connolly, 38–39, Nickolay Stanev, 17, 18, Nynke van Holten, 10, Petros Goulas, 43, r.classen, 40–41, Rainer Lesniewski, 44–45, RENA MICHAEL, 39, RhondaFay, 6, Saran Jantraurai, 21, saraporn, 26, Stavrida, 26, Tomas Ragina, 9, Tracy Burroughs Brown, 23, tusharkoley, 37, U2DO, 18, Vadim 777, 31, Wildnerdpix, 23, www.sandatlas.org, 34; Wikimedia Commons/Q9889, 7 • Printed in China

Library of Congress Cataloging-in-Publication Data Names: Koestler-Grack, Rachel A., 1973- author. | Title: 12 iconic American natural wonders / by Rachel Grack. | Description: Mankato, MN: Black Rabbit Books, [2025] | Series: Iconic America | Includes bibliographical references and index. | Audience: Ages 9–13 | Grades 4–6 | Identifiers: LCCN 2024020325 (print) | ISBN 9781645823971 (library binding) | ISBN 9781645824190 (paperback) | ISBN 9781645824411 (ebook) | Subjects: LCSH: Natural history—North America—Juvenile literature. | Natural monuments—North America—Juvenile literature. | Classification: LCC QH102 .K64 2025 (print) | LCC QH102 (ebook) | DDC 557.3—dc23/eng/20240521 | LC record available at https://lccn.loc.gov/2024020325